ICE WORKS CONSTRUCTION G[UIDES]

Pile Driving
W. A. Dawson

CONTENTS

1.	Introduction	1
2.	Historical	2
3.	Driving theory in practice	4
4.	Driving resistance	5
5.	Sheet piling	10
6.	Pitching piles	15
7.	Driving	17
8.	Pitching and driving sheet steel piles	21
9.	Extracting piles	25
10.	Testing	26
11.	Silent piling	26
Bibliography		27

Thomas Telford Ltd
1981

Other ICE Works Construction Guides available

Earthworks, P. C. Horner *Access*
Scaffolding, C. J. Wilshere

ISBN 0 7277 0093 6

© W. A. Dawson 1981

Published by Thomas Telford Ltd, Telford House, PO Box 101, 26–34 Old Street, London EC1P 1JH

All rights, including translation, reserved. Except for fair copying, no part of this publication may be reproduced, stored in a retrieval system or transmitted in any form or by any means electronic, mechanical, photocopying, recording or otherwise, without the prior written permission of the Managing Editor, Publications Division, Thomas Telford Ltd, PO Box 101, 26–34 Old Street, London EC1P 1JH

Typeset, printed and bound by David Green (Printers) Ltd, Kettering, Northamptonshire

1. Introduction

1.1 This guide note deals with the driving of preformed piles. It excludes the various methods of forming holes in the ground and filling with concrete, collectively known as in situ piles, a speciality requiring a separate note.

1.2 The principal types of preformed piles are timber, precast concrete and steel.

1.3 Definitions

The *hammer* or *ram* is the cast iron or cast steel weight used to drive the piles (Fig. 1).

The *dolly* is a timber or plastic block which is struck by the hammer to 'spread' the kinetic energy of the blow.

The *helmet* is a cast or welded steel cap which fits over the pile head and is socketed on top to hold the dolly.

Leaders are guides fixed to the jib of a crane or excavator, on which the hammer and helmet slide (Fig. 2).

The *piling rig* is the whole assembly: pile frame, helmet, hammer and winch; or crane, leaders, hammer and helmet.

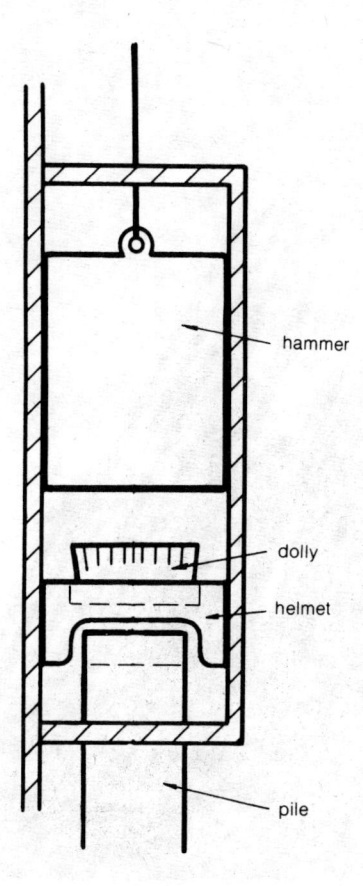

Fig. 1. Hammer, dolly and helmet

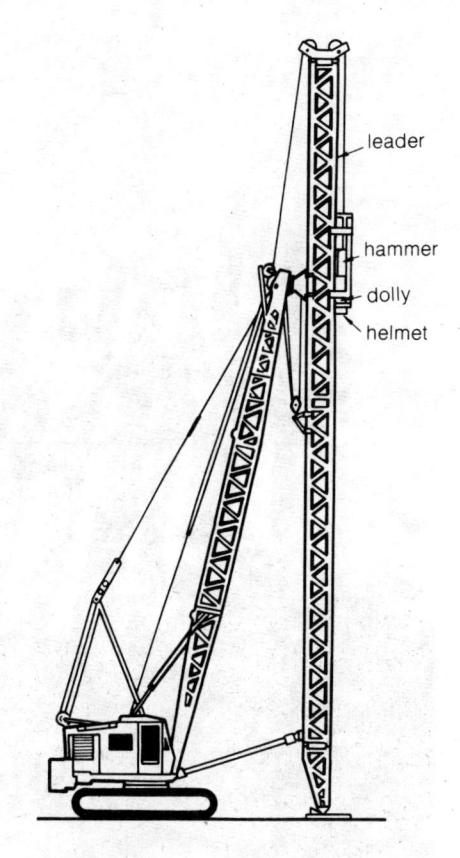

Fig. 2. Modern slideably fixed leader rig

1

Fig. 3. Raking steel piles 35 m long, driven to 450 t resistance

Raking piles are piles driven at a slope away from vertical (Fig. 3).

Pitching a pile is the operation of lifting it from the horizontal or positioning it under the hammer.

2. Historical

2.1 Timber piles have been used for more than ten thousand years. Piled lakeside dwellings are some of the earliest forms of construction. Most classical and medieval riverside towns were built on timber piles and many of our ancient cathedrals are still standing on them. Piling was a recognized military skill as shown by the use of a 'pile' as a heraldic charge (Fig. 4). In the Jewel Tower, opposite the House of Lords, there is a tapered elm pile of

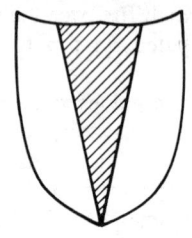

Fig. 4. Pile as a heraldic charge

the shape shown in Fig. 4, recovered from adjoining foundations.

2.2 Timber piles are still probably the most frequently used throughout the world. Many of the American skyscrapers built up to the Second World War stand on timber piles. Long straight pitch-pine trunks make excellent piles. Greenheart is still used in this country for lighter marine structures, and in Australia they have an excellent wood called Turpentine.

2.3 Precast concrete piles were the most popular piles in Great Britain in the middle decades of this century because of their greater bearing capacity than timber, and their cheaper cost than steel. The great drawback of a reinforced concrete pile is its self-weight in relation to its bending resistance when being lifted from the horizontal: the heavy reinforcement found in a reinforced concrete pile is generally to make it possible to lift the pile without it breaking in half.

2.3.1 This latter problem was eased a little by the introduction of *prestressed concrete piles*; prestressed concrete enabled lighter piles to be designed for the same strength. Prestressing however introduced another difficulty. Many piles need shortening or lengthening according to the ground resistance and it is difficult to shorten a prestressed or post-tensioned pile without releasing the stress. Lengthening a precast concrete pile is a slow and costly process, either leaving the pile frame standing while the new concrete hardens, or taking it away to continue driving elsewhere on site and bringing it back later when the new concrete extension is hard enough to drive.

2.3.2 It is because of the difficulties mentioned above that *segmental concrete piles* were introduced. Concrete pile segments are factory-made in standard sizes and standard lengths. The cross-sectional size is selected by the required bearing capacity, and segments of standard length are hauled to the site by any normal transport. They are jointed together by various patented locking devices and additional lengths can be added as driving proceeds. There is an additional advantage: a shorter rig than would be required for a single full-length pile can be used for pitching and driving a segmental pile.

2.4 Steel piles have replaced concrete in many applications since 1960. Experience has shown that the corrosion of steel in the ground is less than supposed and is usually negligible below a water table. Piles have been extracted after 30 years, still showing the rolled-in maker's name. Steel is much more expensive than concrete but the handling, pitching and driving costs are less. A little calculation will show that a steel pile cross-section to carry 400 t is much lighter than a concrete section to carry the same load. The self-weight of the latter in, say, a 25 m long pile makes it a very expensive thing to handle and pitch. The relative labour of driving will appear later. Steel piles can be quickly extended by welding or can be shortened by burning, and the offcuts reused as extensions.

2.4.1 H sections are now most frequently used. A great many sections are available from the mills (see BSC handbooks) and 500 t loadings are not uncommon. The steel H pile is probably superior in hard and/or boulder-infested ground.

2.4.2 Steel tube piles can be supplied in all sizes up to the monster piles supporting the North Sea Oil installations, which are a specialist subject beyond the scope of this note. Steel tube piles fall into two groups: heavy tubes that carry the load on the steel shell; and light tubes that need a concrete filling to carry the load.

Heavy tubes may be circular, as made by all the tube manufacturers, or hexagonal as made by the BSC pile rolling mills. Light tubes may be made by spiral welding, such as supplied by British Steel Piling Co., or light corrugated tube, such as supplied by Armco. These latter may be stepped or tapered.

3. Driving theory in practice

3.1 There is a fundamental relationship that is at the heart of all pile driving which is rarely clearly understood, but which should be fixed in the minds of all piling engineers. A pile hammer hitting a pile is an example of the elementary principle of conservation of momentum. In this application it is as follows.

Weight of hammer (W) × velocity (v_1) = weight of hammer (W) and pile (p) × velocity of the pair (v_2)

that is: $Wv_1 = (W + P)v_2$

but the energy of the hammer = $Wv_1^2/2g$

and $Wv_1^2/2g$ is greater than $(W + P)v_2^2/2g$.

The difference is the loss of energy on impact.

The student should work out the difference between the energy loss when the hammer weight equals the pile weight, and when the hammer weight is half the pile weight. The result will show that *the heavier the hammer the greater the efficiency*.

3.2 This relationship holds good however the hammer is operated. Whether the hammer is lifted by steam, compressed air, diesel expansion or a rope, the efficiency of the blow depends on the hammer weight for any given pile.

3.3 The reverse effect of this is that a lighter pile is more efficiently driven than a heavy pile of the same strength. Calculate the weight of a steel H pile to carry 200 t and a concrete pile to carry the same load, and compare the driving efficiency when using the same-sized hammer.

3.4 In practical terms, on the field one must envisage the temporary elastic compression of the pile, dolly and helmet. When the hammer is too light, nearly all the energy is taken up by the temporary compression, some of which is given back by hammer bounce; and 'set', that is, the penetration per blow, gets smaller and smaller until no progress is made, and because no pile is perfectly elastic the energy being poured in destroys the pile instead of driving it.

3.5 A simple site test (Fig. 5), clearly demonstrating this, is the graphical measurement of the set.

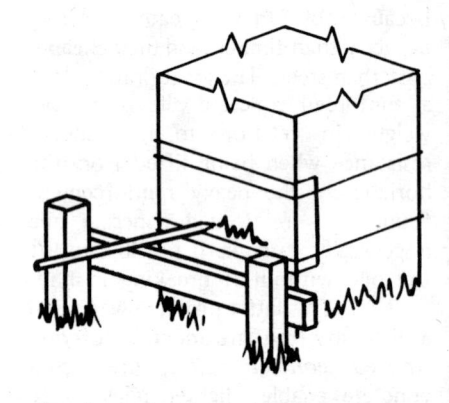

Fig. 5. A simple site test to measure set

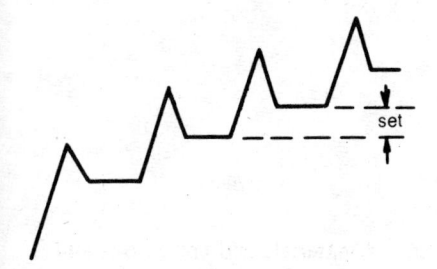

Fig. 6. Typical graph showing set during driving

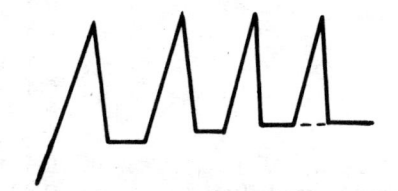

Fig. 7. Typical graph showing refusal during driving

Fasten a sheet of paper on to the pile being driven. Hold a pencil point against the paper (steadied by a horizontal support) and roll it slowly along while the pile is being driven. Fig. 6 shows what the graph produced might look like. The set is the increment of each step. The 'peak' is the elastic compression of the ground and of that part of the pile *below the pencil*.

Too light a hammer, or 'refusal' with an adequate hammer would produce a graph similar to Fig. 7. Refusal is shown by zero set.

As a general rule sets less than 2 mm are to be avoided for continuous driving. Continuous driving at sets of 1 mm almost certainly destroys the pile. If a pile head is collapsing while being driven using a correctly shaped helmet and dolly there is the risk that the pile may be collapsing below ground. Often, when overdriven concrete piles are exposed, it can be seen that they have sheared as shown in Fig. 8. N.B. Wear *your* helmet when doing these tests.

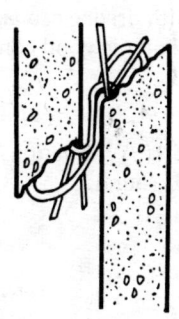

Fig. 8. Sheared overdriven concrete pile

4. Driving resistance

The great advantage of a preformed pile is that its bearing capacity can be calculated from its driving resistance in all but a few exceptional circumstances.

4.1 The Hiley formula gives the simplest method of calculating the driving *resistance* and it models, in mathematical terms, the mechanics of the impact. It is usually written

$$R = (Wh\eta)/(S + C/2)$$

R is the driving resistance
W is the weight of the hammer
h is the fall of the hammer
S is the set per blow (measured in the same units as h)
C is the total elastic compression of pile, dolly and ground
η is the efficiency of the blow

Figures for calculating efficiency of blow are given in Table 1 which is based on Table 7 of the *Civil Engineering Code of Practice No. 4*.

Table 1 clearly illustrates the points made in paragraph 3.1 above, for example
where $P/W = 1/2$, then $\eta = 0.7$
where $P/W = 2$, then $\eta = 0.4$
It might be thought that this would

Table 1(a). Efficiency of blow: value of coefficient of restitution, e

The value of the coefficient of restitution e has been determined experimentally for different materials and conditions and is approximately as follows:

Piles driven with double-acting hammer
    Steel piles without driving cap 0.5
    Reinforced-concrete piles without helmet but with packing on top of pile 0.5
    Reinforced-concrete piles with short dolly in helmet and packing 0.4
    Timber piles 0.4

Piles driven with single-acting and drop hammer
    Reinforced-concrete piles without helmet but with packing on top of piles 0.4
    Steel piles or steel tube of cast-in-place piles fitted with driving cap and short dolly covered by steel plate . . . 0.32
    Reinforced-concrete piles with helmet and packing, dolly in good condition 0.25
    Timber piles in good condition 0.25
    Timber piles in poor condition 0.0

The efficiency of the blow can be obtained from Table 1(b) for various combinations of e with the ratio P/W, provided that W is greater than Pe and the piles are driven into penetrable ground.

Table 1(b). Calculations of efficiency of blow, η

P/W	$e = 0.5$	$e = 0.4$	$e = 0.32$	$e = 0.25$	$e = 0.0$
½	0.75	0.72	0.70	0.69	0.67
1	0.63	0.58	0.55	0.53	0.50
1½	0.55	0.50	0.46	0.44	0.40
2	0.50	0.44	0.40	0.37	0.33
2½	0.45	0.40	0.36	0.33	0.28
3	0.42	0.36	0.33	0.30	0.25
4	0.36	0.31	0.28	0.25	0.20
5	0.31	0.27	0.25	0.21	0.16
6	0.27	0.24	0.23	0.19	0.14

Table 2. Temporary compressions in mm

Form of compression	Material	Easy driving	Medium driving	Hard driving	Very hard driving
Pile head and cap: C_c	Head of timber pile.	1	2·5	4	5
	Short dolly in helmet or driving cap.*	1	2·5	4	5
	75mm packing under helmet or driving cap.*	2	4	5·5	7·5
	25mm pad only on head of reinforced concrete pile.	1	1	2	2·5
Pile length: C_p	Timber pile. E: 10.5×10^3 N/mm²	$0.33\,L$†	$0.67\,L$†	$1.00\,L$†	$1.33\,L$†
	Pre-cast concrete pile. E: 14.0×10^2 N/mm²	$0.25\,L$†	$0.5\,L$†	$0.75\,L$†	$1.00\,L$†
	Steel pile, steel tube, or steel mandrel for cast-in-place pile. E: 210×10^2 N/mm²	$0.25\,L$†	$0.5\,L$†	$0.75\,L$†	$1.00\,L$†
Quake: C_q	Ground surrounding pile and under pile point.	1	·2·5 to 5	4 to 6	1 to 4

*If these devices are used in combination, the compressions should be added together
†Length L measured in metres

increase the driving time by 7/4, a serious enough increase, but the effect of the temporary compression makes it much worse than that.

Temporary compressions for different grades of driving are given in Table 2 which is based on Table 8 of the *Civil Engineering Code of Practice No. 4*. The comparative hardness of driving is usually classified in terms of the compressive stress per unit of cross-sectional area of pile, or shoe (Table 3).

Table 3

Easy driving	3·5 N/mm^2
Medium driving	7·0
Hard driving	10·5
Very hard driving	14·0

In steel piles, only the cross-sectional area of the steel is taken into account and the four respective driving stresses are taken as 50, 100, 150 and 200 N/mm^2.

4.2 The Hiley formula can be re-written

$$Wh\eta = R(S + C/2)$$

i.e. the energy in the falling hammer X η = the distance moved by the pile X the resistance to driving.

When making use of the formula, you will know $Wh\eta$. Select the resistance you are aiming for. The value of $(S + C/2)$ can be found; it may be, say, 7 mm. However, when you look up the temporary compression (Table 2), you may get a value of, say, 12 mm. Taking $C/2$ from the calculated value of $(S + C/2)$ leaves a set of only 1 mm. This means that *you are using too light a hammer.*

4.3 In effect the temporary compression represents a fixed amount of energy (for a given value of R) which has to be expended before the remaining energy of the blow moves the pile downwards. So with too light a hammer you can be going nowhere fast or at least very slowly.

4.4 Do not think that a lack of energy can be overcome by increasing the hammer drop. Falls of much over 1½ m for continuous periods will destroy pile and piling equipment, except on certain of the larger diesel and steam hammers in the high price range.

4.5 There is of course a practical maximum hammer weight for any given pile, which may be calculated by the driving stress on the head of the pile (*Civil Engineers Code of Practice No. 4*, paragraph 3.83). If, after running through the set calculations, you arrive at a hammer weight too big for the pile it indicates that someone has chosen the wrong size of pile.

4.6 Stresses in continuous hard driving. The most difficult and costly problem for the piling engineer is driving through hard ground or a series of hard beds that will not quite take the bearing load. This results in long periods, sometimes days, of driving at sets which do not quite reach the desired minimum for the load to be carried by the pile. It may be that the driving stresses are less than maximum; nevertheless, a continuous amount of energy is being poured in, which is not being used up by the forward progress or given back by the rebound. This can result in fatigue and collapse of the pile — not always where it can be seen.

4.7 Where continuous hard driving is anticipated, to get through hard beds of some thickness, a stronger pile should be selected. To try to save money by using a lighter pile is to risk

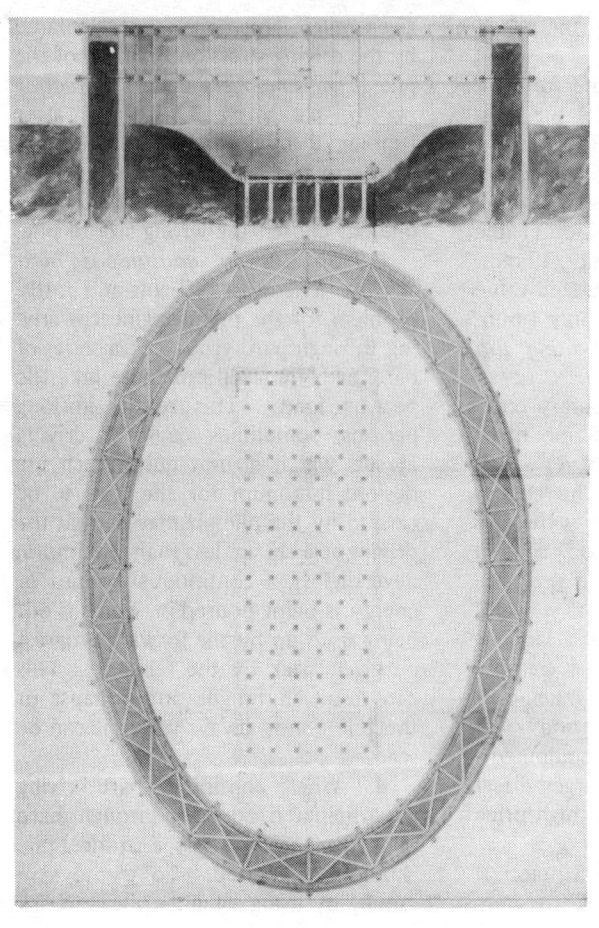

Fig. 9. Plan of Rennie's timber cofferdam

an inferior foundation, delays, disputes and perhaps a costly change of design in the end. Remember that soil investigation reports can only be approximate.

5. Sheet piling
5.1 Historical
The use of piles driven side by side to form a cofferdam goes back more than two thousand years. The most striking example early in English history is the construction of the medieval London Bridge in the reign of King John, in about 1215. The massive boat-shaped masonry foundations had to be built within timber cofferdams. Anyone who saw the savage tidal flow between the steel cofferdams when the new London Bridge was built recently, can only have amazed respect for the men who built the much greater obstructions to flow of the medieval bridge within timber cofferdams. When Rennie designed the nineteenth century London Bridge he also designed the timber cofferdams within which the foundations were constructed (Fig. 9).

5.2 Timber sheet piles could be tongued and grooved like a very thick floor board; for any depth of water it was necessary to use a double skin with clay filling between the skins, of sufficient thickness to form a gravity dam. They are now rarely used.

5.3 Interlocking steel sheet piles were first used regularly between the First and Second World Wars, and have been used almost universally since 1945. In the 30 years since 1950 the use of steel sheet piles in Great Britain has increased from 50 000 t per annum to 200 000 t per annum.

A large variety of sections are rolled in the two types, Larssen and Appleby Frodingham, and in lengths exceeding 40 m in the heaviest sections. The lighter sections are used in excavations that once would have been timbered. The long, heavy sections ease such problems as building cofferdams for massive foundations in deep tidal water, like those for the Forth, Severn and Humber Suspension Bridges.

Steel sheet piles are also used increasingly in permanent retaining structures; they are prefabricated units that can be quickly installed, a valuable advantage in earth-retaining problems (Figs. 10-13).

Fig. 10. Two steel sheet pile walls tied together contain a road embankment (courtesy of Hoesch, Estel, Huttenverkaufskontor GMBH)

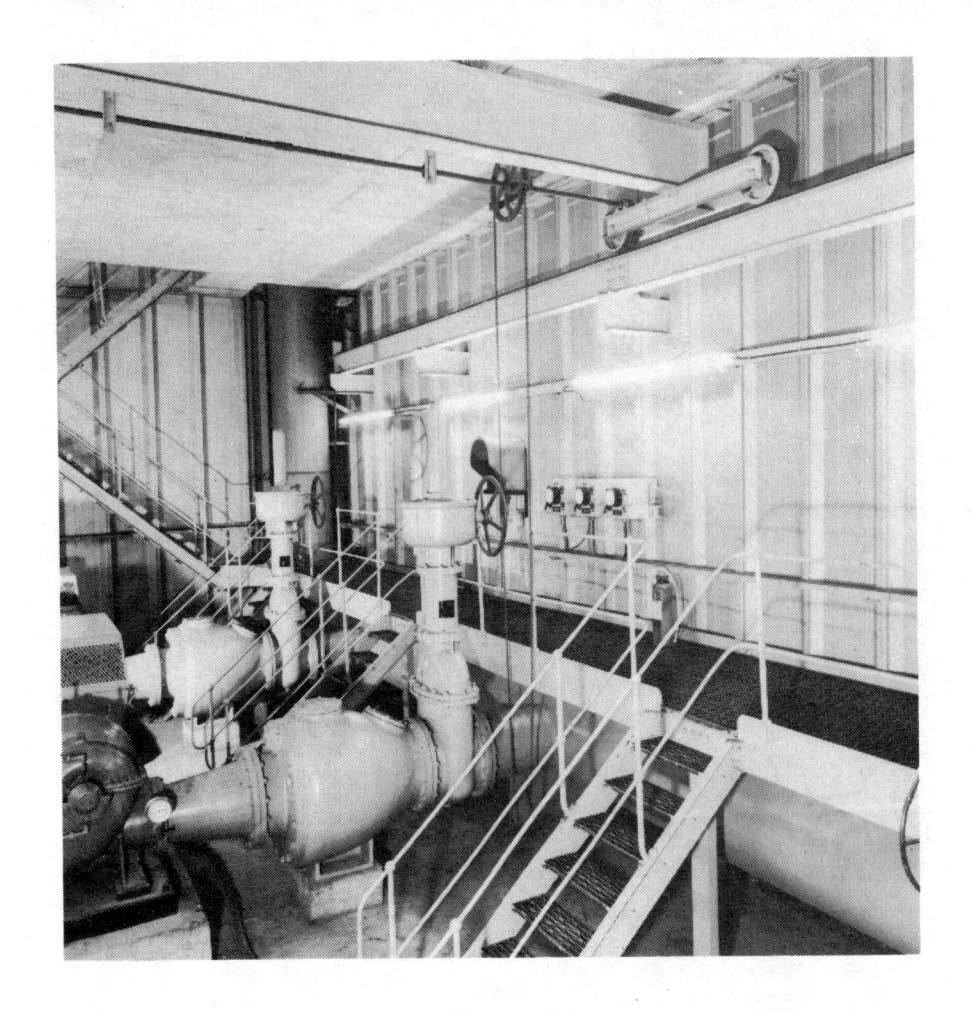

Fig. 11. Unissen piles used as a permanent wall (courtesy of British Steel Corporation)

Fig. 12. Use of steel sheet piling keeps the building site very small and reduces construction period considerably. To obtain the curve in the road line the piles were slightly turned in the interlocks without impairment to the driving process (courtesy of Hoesch, Estel, Huttenwerkauf-skontor GMBH)

Fig. 13. Sheet steel piling, beside fast mounting, offers the opportunity of building economically wing walls that meet the demands of landscape designers (courtesy Hoesch, Estel, Huttenverkaufskontor GMBH)

Familiarity with the types of sheet piles and their properties can be gained by obtaining the handbooks of the British Steel Corporation.

6. Pitching piles

The importance of pitching piles is often neglected. In some cases, such as the pitching and securing, in correct position and rake, of long heavy piles in tidal water, the pitching is a task which can take three times as long or more as the actual driving. In fact, in very many cases, it is the pitching problem which governs the height of the rig used.

6.1 Pitching reinforced concrete piles involves special care because of their large self-weight in relation to their bending resistance; for piles over 10 m in length it is necessary to pick them up at their 1/5th points (in order to minimize bending due to self-weight) using special holes cast in (Fig. 14). This requires a crane with two independent winch drums in order to ensure a controlled pitching.

6.2 Timber piles are usually relatively short (10 m) lengths and they can be lifted from their top ends.

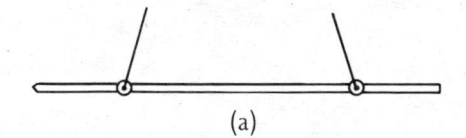

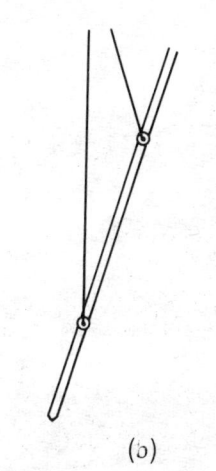

Fig. 14. Reinforced concrete pile, picked up at 1/5th points: (a) lifted and (b) pitched

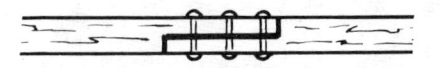

Fig. 15. Long timber pile formed by splicing

Sometimes longer timber piles have to be formed by splicing (Fig. 15). Then it is necessary to lift at the 1/5th points, as with a concrete pile, using a two-drum crane. Alternatively the first half can be driven and the second half spliced on in situ.

6.3 It would be an unusual steel pile which could not be lifted (with care) from one end (Figs. 16, 17).

6.4 Having got your pile up on end, against leaders or held by temporary works, the lifting rope (or ropes) has to be disconnected. In the days of piling frames men used to climb up sometimes 25 or 30 m to struggle with a heavy bolt, but today piling frames are rarely used and there is nothing to climb on. This essential operation is performed by a 'quick release shackle' operated by a cord from the ground.

If the quick release shackle does not work you have to bring the pile back down again. This may be beyond the power of the crane when a heavy pile has sunk deeply into the mud. It is essential, therefore, to keep quick release shackles (and spares) in good repair, or the young engineer might find himself doing a heroic

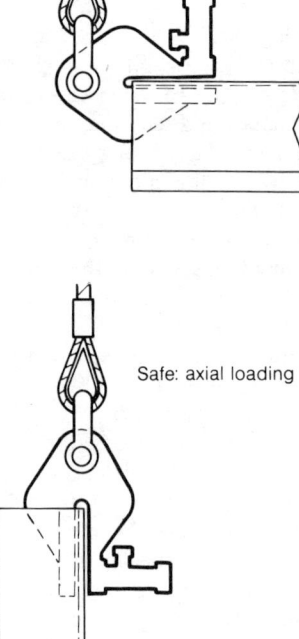

Fig. 16. Steel pile being lifted from one end

Fig. 17. Safe methods of pile lifting

16

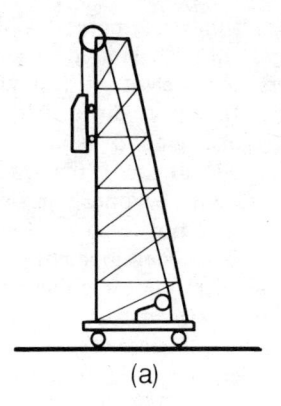

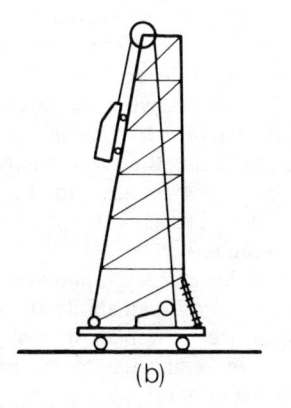

Fig. 18. Drop hammer mounted on frame: (a) vertical and (b) raked

task, clinging on to a crane jib with one hand over 50 m above the ground, trying to release a jammed shackle. (It is always blowing hard, wet and freezing when this happens.)

7. Driving

The pile hammers most generally used today are drop hammers and diesel hammers. Single- and double-acting steam or compressed air hammers are less frequently used, except for the ultra-large single-acting hammers found on offshore oil platforms. Pile hammers should be designed on the principle that all hammers shake themselves to bits, so the fewer bits they are made of, the better.

7.1 Drop hammers are simply large weights in cast iron (or cast steel for special shapes). Until the 1930s they were nearly always operated by a winch (steam, petrol or diesel) and mounted on a tall frame which in some cases could be raked (Fig. 18). The frames were usually mounted on rollers and needed a gang of labourers to move them between piles.

Sometimes a pile could be pitched with the frame and winch if the piles were stacked close to the frame, but it was easy to pull it over and, for a reasonable rate of production, a separate crane was needed for pile unloading and pitching. Frames always needed guying and the guys needed moving and adjusting for every pile, a tedious and labour intensive task.

7.1.2.1 In the early 1930s excavator manufacturers began to offer pile leaders with their basic machines (Fig. 19). These were pinned at the

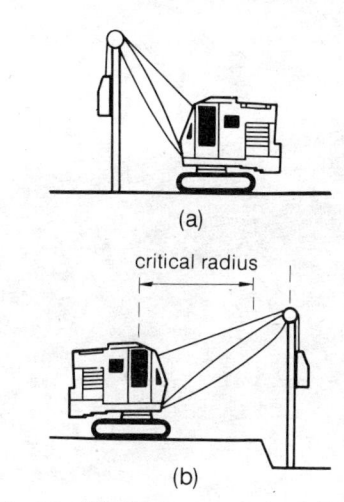

Fig. 19. Pile leader on crane; (b) shows critical radius

Fig. 20. Crane-leader piling rig with slideable leaders

jibhead and were more flexible than frames, although limited in length of leaders and weight of hammer. In practice they were difficult to operate on irregular ground as they had to derrick out to lower the leader and were quickly beyond their working radius (Fig. 19(b)).

7.1.2.2 A development of the crane–leader type of piling rig, often specially designed and built by piling contractors themselves, is to have the leaders slideably-fixed to the jibhead (Fig. 20). This gives great flexibility both in rake and acceptable level of pile foot. To obtain the maximum use of a leader rig the crane should be able to swing round to pick up its own pile. This means that at the desired radius the crane has to lift the leader, the hammer and the pile, and it will be found that a substantial crane is required.

7.1.2.3 A final development of the leader type of rig is an ability to pivot the leader (seen in plan in Fig. 21), otherwise the crane has to reposition itself for every pile.

The most flexible arrangement for drop hammers is the rope-suspended drop hammer working within its own frame and carrying its own helmet and dolly (Fig. 22). This hammer can drive any pattern of piles within the radius of the crane without moving the crane (Fig. 23).

Its limitations are that it requires upper and lower walings to hold the piles (where these are in rows) (Fig. 24) or prefabricated tripods for single and irregularly spaced piles (Fig. 25). A rope-suspended drop hammer cannot be used with raking concrete piles, because of their low resistance to breakage. However this type of hammer is particularly useful and flexible when driving piles for marine structures over water.

7.2 Diesel hammers

In applications the choice of diesel hammer or rope-suspended drop

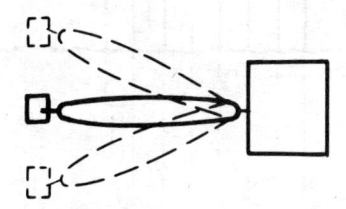

Fig. 21. Pivoting of the leader seen in plan

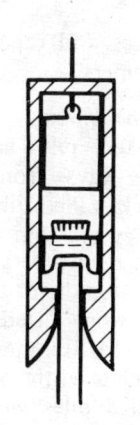

Fig. 22. Rope-suspended drop hammer working within its own frame

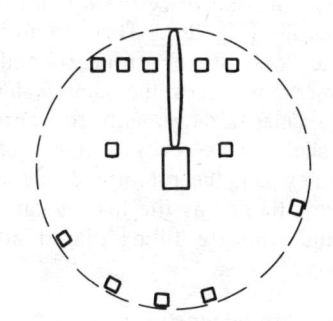

Fig. 23. Piles within the radius of the leader can be driven without the crane being moved

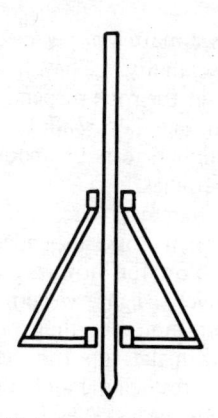

Fig. 24. Waling to hold row of piles

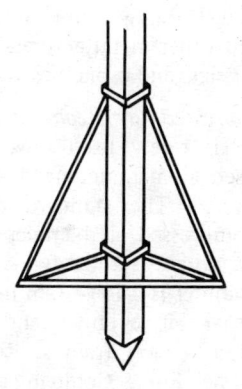

Fig. 25. Tripod to hold single pile

19

hammer is a matter of personal choice and/or availability. They have equal flexibility in the rope-suspended mode and need similar waling systems. Diesel hammers can be mounted on leaders or frames.

Diesel hammers are usually less efficient than rope-suspended drop-hammers, blow for blow, as a result of the relatively light weight of the hammer or ram; but their rapid rate of striking makes up for the lighter blow, by producing more blows per minute. A heavy diesel hammer is superior to a rope-suspended drop hammer where long periods of hard driving are encountered. Refer to British Steel Piling Company's, Delmag's and other manufacturers' handbooks for sizes and capacities available.

7.3 Compressed air/steam hammers

Steam is rarely used now and a compressed air hammer needs a large compressor. The standard double-acting compressed air hammer is very noisy by modern standards. A single-acting hammer is a large ram lifted by an internal air cylinder and needs heavy leaders or a frame. Refer to British Steel Piling Company's handbook for sizes and capacities of air operated hammers.

7.4 Internal hammers

Tubular steel piles are driven in two separate ways: from the top if they are thick walled tubes; from the bottom if they are thin walled tubes.

All normal hammers listed previously can top-drive thick walled tubular piles when fitted with a suitable helmet to fit the internal or external diameter of the pile.

Thin walled piles would collapse if top-driven and they have to be driven by an internal cylindrical hammer driving on to a plug of gravel or dry concrete placed in the bottom of the pile (Fig. 26). This system has all the flexibility of a rope-suspended hammer and needs the same walings or guidance arrangements to control the pile. It is really a transitional type between the preformed pile and the in situ pile as the load is carried on the concrete filling placed after driving.

7.5 Cost of hammers

The cost of a rope-suspended hammer is a fraction of the cost of

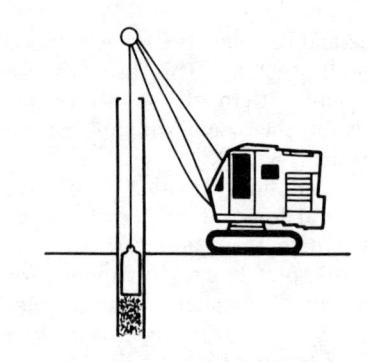

Fig. 26. Thin walled pile driven by internal hammer

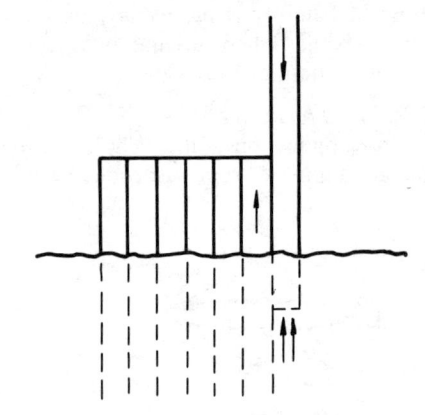

Fig. 27. Asymmetrical resistance to driving in steel sheet piles

20

sophisticated leader equipment or a diesel hammer of similar energy. It is not a fact of pile driving that double the cost of hammer can be paid for by double the driving rate because of the time spent pitching. The hammer is standing idle while the pile is being pitched (see paragraph 6).

8. Pitching and driving steel sheet piles

8.1 The special problem in driving steel sheet piles is that the drag on the interlock makes the resistance to driving asymmetrical (Fig. 27). This results in a panel of piles leaning forward (Fig. 28).

A means of partially reducing this effect is to pitch and drive the piles in pairs. The best remedy however, is to pitch piles in panels and drive in reverse order (Fig. 29). In order to achieve this a crane which can interlock a pile on top of the last undriven pile is needed. This is often the governing factor in the selection of crane size (Fig. 30). For instance sometimes a 40 m pile has to be threaded on to another 40 m pile.

Steel sheet piles can be supplied run together in pairs.

8.2 Steel sheet piles which are not

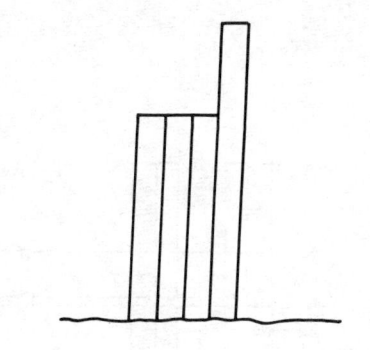

Fig. 28. Asymmetrical resistance driving resulting in forward lean

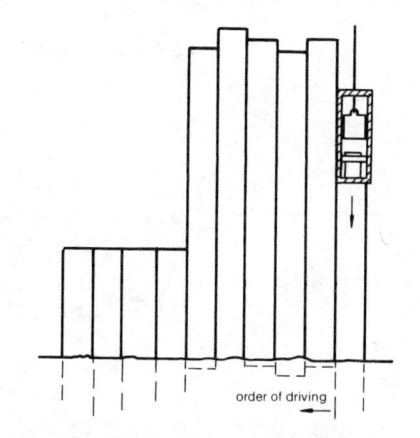

Fig. 29. Reverse order of drive to prevent forward lean

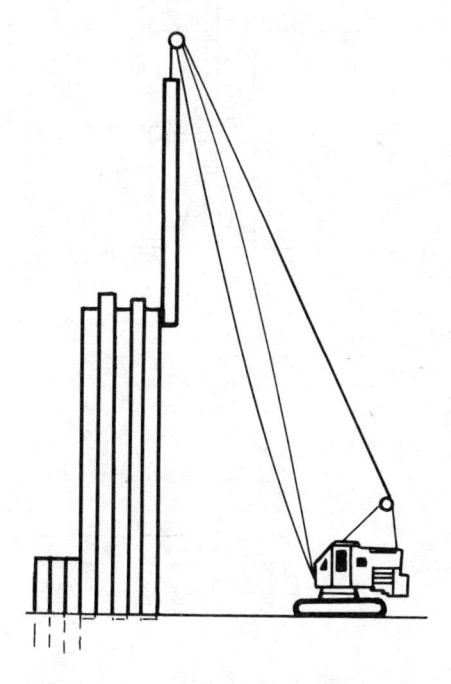

Fig. 30. Threading of one pile on to another may determine crane size

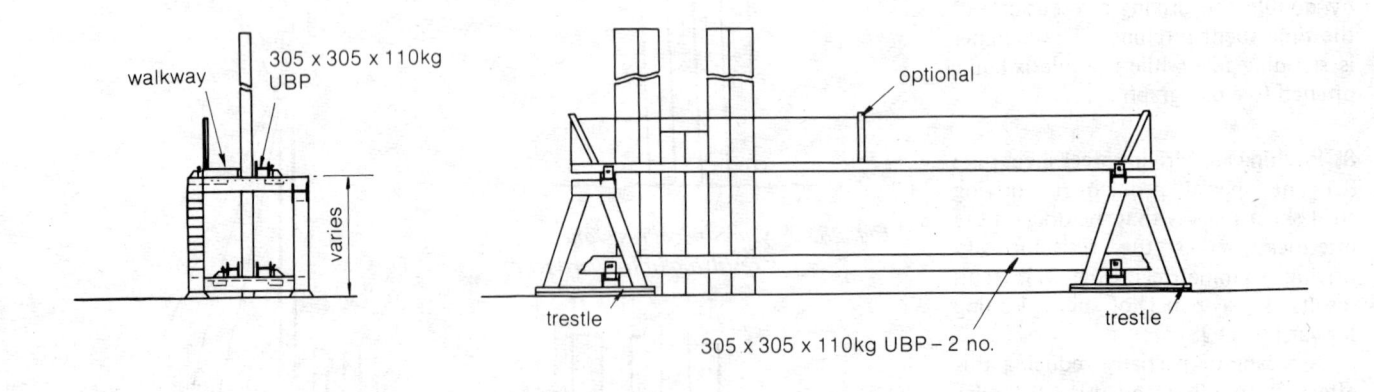

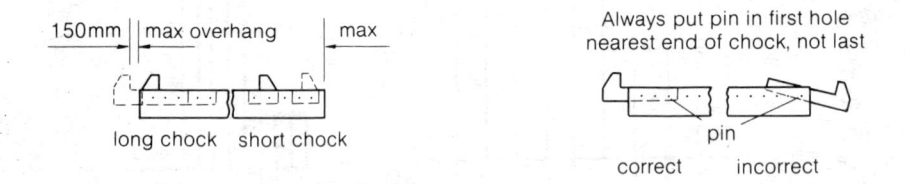

Fig. 31. Normal erection of walkway waling using trestles both ends

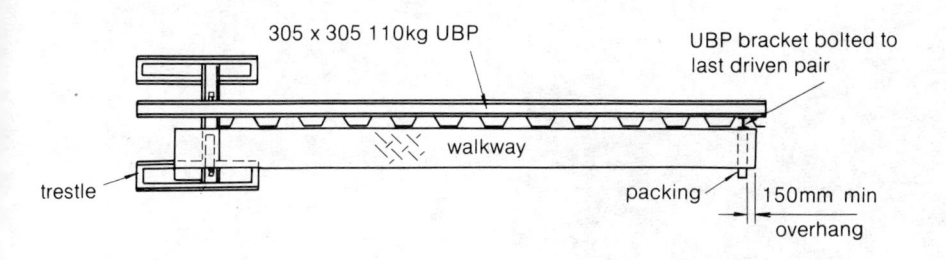

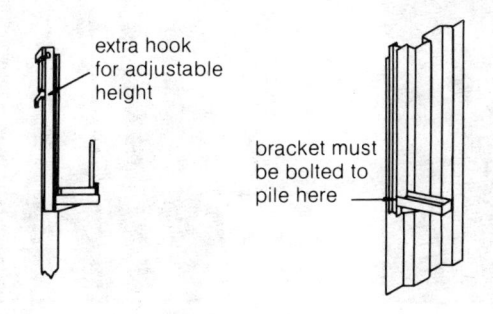

Fig. 32. Normal erection of walkway waling using trestle one end, hanger bracket the other

driven singly by frame or leaders need a waling system for guidance, which should have a safe walkway. These walings are often heavy timbers bolted up on site, but there is a modern tendency to provide purpose-made frames for repeated use. Where continuing a line of piles already driven, the ends of the walkway walings are hung on the last pile driven (Figs. 31, 32).

8.3 Threading the interlocks

This is a difficult and dangerous task in which the interlock of the pile being pitched has to be threaded into that of the immediately preceding pile. Standing perhaps 30 m above the ground, the workman has to handle something weighing five or six tonnes, swaying at the end of a long jib, and enter it into a slot with only 1 mm clearance. Many fingers and sometimes lives have been lost doing this.

This hazardous operation can be eliminated by the use of a device known as a pile threader (Fig. 33). This is fastened onto the pile being lifted into position, about 1 m from the bottom. The jib lifts the new pile up to the top of the sheet piling,

23

Fig. 33. Pile threader attached to pile; inset: pile threader

Fig. 34. An extreme case of how not to thread piles

Fig. 35. Method of operating the pile threader; the operator works either from ground level or from top staging

guided by rollers, where the threader automatically dovetails the interlock of the new pile into that of the last one pitched, and slides the new pile down into place, thus eliminating the need to have an operative at the top of the piles (Figs. 34, 35).

8.4 Hammers for steel sheet piles

The same hammers are used on driving steel sheet piles as on other steel piles, but there is the additional problem that the hammer width must not be wider than a single pile or a pair of piles, whichever is being driven. It is essential to have a helmet in the hammer which accurately fits the section of pile being driven.

8.5 There are two additional methods of driving, mostly applied to sheet piles. These are the vibrator and the hydraulic pile driver.

8.5.1 Vibrators. These have been developed in increasing weight and power and the power has to be supplied by a mobile generator of the order of 150 kw. Vibrators can be miraculous in some types of granular material and unable to drive at all in some types of cohesive material. Refer to Tomen and Schenke catalogues for details of sizes and power required.

Hydraulic pile driver. This is a pack of hydraulic rams, spaced at the same centres as the piles to be driven. The rams hold on to five piles to obtain a reaction and thrust down on the sixth. In this way they work down each pile in turn. The driving is almost completely silent but a mobile power unit is needed to supply the hydraulic pressure. This method is excellent in cohesive ground but can have difficulty in some granular material and obviously it can only be used for driving piles in groups.

9. Extracting piles

9.1 Extraction of piles is frequently required so that steel sheet piles used as temporary support can be recovered for further reuse. These can often be used many times before they become too short as a result of repeated cutting to a new length and cropping off of damaged tops. Efficient extraction is therefore an important economic factor in the use of steel sheet piling.

Many attempts are made to develop pile pullers by multireeved blocks and a strut under the crane jib head, but this type of equipment has a very limited range of use.

9.2 Compressed air extractors have been used for many years and still are masters in the field. Size for size you choose from the maker's or hirer's lists.

You must have a crane capable of exerting the maximum pull specified for the extractor, as a large extractor on a light crane merely shakes the jib to pieces, and this pull must be for the radius at which the extractor will be required to operate. You must have a compressor with a capacity at least that specified for the extractor selected, to ensure that it can give it air flow at full pressure. You are wasting your time with an underpowered compressor. Recent developments in extraction practice now go up to 1.25 N/mm^2. Refer to British Steel Piling Company's catalogues for sizes of extractors available.

The only competition to the compressed air extractor is the more recently developed vibrator mentioned in paragraph 8.5.1. In some types of ground it can work much better than the compressed air extractor. It is very difficult to forecast which equipment will be more effective. Try one type, and if the results are disappointing try the other. Remember that it is essential to have an equally powerful crane for a vibrator as for a compressed air extractor.

10. Testing
Refer to *'Piling; model procedures and specifications'* for a description of the various tests which may be required.

Where you are required to draft a specification for pile testing, you should bear in mind the high cost of pile tests. For instance if Kentledge has to be brought to the site, set up, dismantled and hauled away the cost for this alone will be over £1000. The test procedure in Table 8 of *Piling model procedures and specifications* is a 24 hour cycle, usually with expensive piling equipment standing on site waiting.

It is not unknown for the pile testing on a small piling contract to cost more than the piles. There have been cases where the young enthusiast has specified a series of tests with extensive loading and unloading restrictions, and the time for these added together was greater than the piling contract time!

11. Silent piling
The expression is almost a contradiction in terms, but much effort has been put into noise reduction, principally by contractors developing their own special equipment.

Various operators have developed various methods of shrouding the hammer to reduce noise, and efforts in other directions have been made to introduce cushioning material into the dolly without reducing the impact too much. Shrouding arrangements make the equipment more cumbersome and therefore slower to use. Cushioning makes the blow less effective and more blows are required.

Quiet piling is more expensive — sometimes by a factor of 5 — and the lesser 'pollution of the environment' is continued for a longer period. It is sometimes found that residents prefer the loud noise of vigorous driving for a shorter period to longer periods of 'silent piling'. Refer to the technical press for those firms advertising quiet or silent piling.

Bibliography

1. G. Cornfield. *Steel bearing piles.* Constrado: Croydon. Reprinting.
2. T. Whitaker. *The design of piled foundations.* Pergamon Press: Oxford. 1976.
3. *British Steel Corporation Piling Handbook 1976.* British Steel Corporation.
4. *Civil Engineering Code of Practice No. 4: Foundations.* Institution of Civil Engineers: London. 1954.
5. *Code of Practice 2004.* British Standards Institution: London. 1972.
6. *Piling; model procedures and specifications.* Institution of Civil Engineers: London. 1978.
7. *Handbooks on Piling Equipment.* British Steel Piling Co.: Claydon, Suffolk.